AF812434

MATH UNIVERSE:

EPISODES IN A SPACE OPERA

[PART 1]

PLAYS BY MANDY KHOSHNEVISAN

ROYA
PUBLICATIONS

For information:

Email: info@royapublications.com • Visit: www.royapublications.com

Cover design by Mandana Khoshnevisan

ISBN: 978-1-944218-09-6 (81/2 x 11 Paperback)

First Edition: September 2021

10 9 8 7 6 5 4 3 2 1

MATH UNIVERSE:

EPISODES IN A SPACE OPERA

[PART 1]

PLAYS BY MANDY KHOSHNEVISAN

Life is difficult on the fringe worlds outside the Empire…
but every day, those who seek truth, knowledge, and justice for the
oppressed push on — living on remote planets,
encountering yet-unknown life forms,
and solving their problems with their
most important tool: MATH.

A Note by the Author:

This collection is the first of a series of short play collections, written with STEM curriculum standards in mind — AND with a cheeky sense of humor that will hopefully appeal to teachers and students alike. These plays are for kids to act out (in person, or remotely/online!), and the action in each helps explain fundamental math and science concepts through metaphor, character interaction, and comedy.

The plays are short, and each collection is connected thematically, so that the individual plays can be performed either alone or as a group in any combination. All characters can be played by anyone of any gender. The characters also have a wide range of dialogue requirements, ranging from verbose characters with multi-sentence lines to read or memorize, to group roles where characters speak in unison and/or contribute with nonverbal communication. The plays could be memorized, or performed as a reading, with scripts in hand. Stage directions could also be read aloud by a stage manager or narrator figure, especially if the play is performed with students being remote. (Students who might not be terribly keen to perform are often excited about being stage managers, which is also an excellent chance to let them show responsibility and is something very helpful for directors: stage managers can take blocking notes and other notes for the director and actors, can be on book for rehearsals, can give actors cues to begin, can run lights or sound effects, and — in person — can take charge of set pieces and props.)

At the end of the collection is a Teacher Planning Section, with handy references for each play, summarizing the mathematical concepts presented, and listing a breakdown of the roles in each play: approximate line counts and short character descriptions, pointing out which characters are more verbal and which ones use more physical acting and few to no words — as a guide for casting.

Though the main action of these mini-stories is solved by teaching math concepts, the characters also have other objectives and they have relationships with each other — they may be new friends or old friends, they may be competing with each other, they may be trying to impress the boss, etc — so hopefully there is fodder for actors and directors to have some fun with characterization and Broad Acting Moments. In any play, the main action is often just an excuse to see characters interact with each other, so even though we're talking about math here, there are clues to how the characters feel about each other and about life in general. There are several plays in this collection that feature the same group of rebels, and though they are identified by number, they were written to be the same characters across all the plays; they have the same attitudes and tendencies — although there's NO reason they couldn't all be played by different student actors in performance. (A fun exercise might also be to have students write their own plays in the same universe with the same or similar characters, working in their own math concepts!)

I sincerely hope you and your students have fun with these plays! May the math be with you ...

—Mandy Khoshnevisan

TABLE OF CONTENTS

STAGE VOCABULARY

For teachers who may not be drama teachers, or students who may not be accustomed to acting or directing, here's some quick Stage Vocabulary. It really comes in handy to have everyone using the same vocabulary when giving directions on how to move around! I've used the same terms when working online as in person, assuming that everyone's camera is the "audience," and therefore "downstage." (And hey! Using words to describe an actor's location and movement is also kind of like math, right?)

Blocking: A term for an actor's movements and traffic pattern during a scene, as related to the timing of the action and dialogue. Where they come in, where they walk, whether they sit or stand, etc. Actors often write their blocking into their script, so they can remember it; and a stage manager will often take blocking notes for everyone, to help the actors remember where to go and what to do once they're **off book** and no longer looking at their script. As in, "Oops! I went out the wrong door, because I forgot my blocking!"

Off Book: Someone who is "off book" has their lines memorized and is working without reading from the script. As in, "Hey there, can you help run lines with me so that I can get off book?"

On Book: Someone who is "on book" is looking at the script. When the actors are rehearsing without scripts in hands, the stage manager is often "on book" for them, to make sure no one forgets anything or to feed them a line if they get stuck. As in, "Let's run the scene from the top — stage management, can you please be on book for us?"

Enter: To arrive somehow on the stage. As in, "When you enter, you're coming from outside, so be sure to show us how cold it is out there!"

Exit: To leave the stage. As in, "At the end, wave goodbye to your friend, and then exit through the upstage door."

Wings: If your stage is set up this way, it might have "wings," or places to wait on the sides of the stage before you enter. Usually the wings have curtains or flats to hide actors from the audience — but even if you don't, we assume that if you're "in the wings," we can't see you.

Downstage: The edge of the stage closest to the audience, OR moving in that direction. As in, "The window is on the invisible downstage wall, so when you're checking outside for monsters, come all the way downstage and peek through the imaginary curtains."

Stage Right: When you're onstage facing the audience (aka, facing "downstage"), the "stage right" wall is the one to *your* right. That direction is always called stage right, no matter which way an individual actor is facing.

Stage Left: When you're onstage facing the audience (aka, facing "downstage"), the "stage left" wall will be on *your* left. That direction is always called stage left, no matter which way an actor is facing.

Center Stage: The very middle of the stage; also used frequently in combination with another description, as in, "When you realize you forgot to unplug the iron, you can come all the way from upstage center to downstage center."

Upstage: Opposite of downstage, and therefore — away from the audience! As in, "If you keep walking upstage, eventually you'll be walking up the wall!"

(You can also use a combination of these directions to further describe locations or directions of movement: "upstage right," "downstage left," "upstage center," etc.)

Onstage: You're on the stage, yes! But also, used as a direction meaning, "sideways, more towards the center." As in, "Can you stand onstage of that tree?"

Offstage: You're off the stage, yes! But also, used as a direction meaning, "sideways, farther from the center," as in, "When you're trying to hide from the monsters, stand on the offstage side of that tree so they don't see you."

Prop: A prop is any object that an actor might use or carry as part of their action.

Set: The "set" is made up of all the furniture, walls, doors, and other elements that define and describe the location. An individual part of the set is called a "set piece," and if you move the set pieces around to create a new location for a new scene, that's called a "set change." (Separate objects, like posters or plants or other things that actors don't need to interact with, whose purpose is just to decorate the set and make it look like a real place, are called "set dressing.")

Preset: Something that has to be onstage before the scene starts is "preset" or "a preset." As in, "Stage management, can you preset the breakroom table, and the props that go on it, please?"

Lights Up/Lights Down: "Lights up" = the theater lights turn on; "Lights down" = the theater lights turn off. Now that most theaters don't use their curtains, lights up/lights down is a good way to indicate when a scene begins and ends. Depending on how formal a presentation it is, you might turn the actual lights on and off where indicated, or just have someone SAY lights up/lights down to help the audience imagine that it's happening. As in, "Actors are preset onstage, and ... LIGHTS UP!"

REBEL BASE 10

by Mandy Khoshnevisan

MATH CONCEPTS:

Estimating; number place value; counting in Base 10

CHARACTERS:

COMMANDER 1: A Commander of the Rebel forces — the one in charge

COMMANDER 2: A Commander of the Rebel forces — second in command

3: On patrol squad A

4: On patrol squad A

5: On patrol squad A — not the quickest thinker, perhaps, but surprisingly gifted in math concepts

6: On patrol squad B

7: On patrol squad B

8: On patrol squad B

9: On patrol squad B

10: The Pilot

PROPS/NOTES:

No props or furniture specifically called for; feel free to use your imagination!

SETTING:

We are in the meeting room of Rebel Base 10, somewhere a long while ago in a galaxy that's certainly not very close. There may well be Wars between Stars out there, is what I'm saying.

Lights up: 1 and 2 are pacing back and forth, looking worried.

They walk to the ends of the stage, they pause, they walk past each other to the other ends, they pause. They walk back towards each other to meet in the middle. They look at each other and shrug and shake their heads.

3, 4, 5 run in.

3, 4 & 5:

Three, Four, and Five reporting for duty! *(They stand in a little line and salute)*

COMMANDER 1:

At last! Where have you been?

3:

Out patrolling, commander.

4:

We didn't get lost, commander.

3:

Not lost at all.

4:

Nope! We knew where we were the whoooooole time.

5:

But you guys said--

3 and 4:

SHHHHHHHHHHH!

5:

Oh yeah! Riiiiiiiiight. DON'T tell the commander we got lost.

3 and 4 face palm

COMMANDER 2:

Well?!? Where's everyone else?

6, 7, 8, and 9 run in

COMMANDER 2:

Where have YOU been?

6:

NOT getting ice cream, commander!

7, 8 and 9 shake their heads and also face palm

COMMANDER 2:

Okay, okay, we get it. That's enough.

3, 4, 5, 6, 7, 8, 9 scurry around and form a line in order of their number, and stand at attention

COMMANDER 1:

Now: does everyone know why I called you in here today?

3:

I do. Don't you, Number Four?

4:

Definitely, Number Three.

COMMANDER 1:

Really? Then why don't you tell us all?

3:

I would, commander, but… I thought it would… be… rude?

COMMANDER 1:

All right, all right. Because I'm feeling generous, I'll just *remind* you, shall I?

All the others nod with relief.

COMMANDER 1: (continued)

Number Two, why are YOU nodding?

COMMANDER 2:

I… am just looking forward to… hearing you explain it, commander?

COMMANDER 1:

The COUNT!!

All the others jump in surprise at the loud sound.

COMMANDER 1: ***(Repeated, whispering so as not to scare everyone this time)***

…the count!

2-9: ***(also whispering)***

Ohhhhhh right, the count!

6: ***(to 5)***

The count?

5: ***(to 6)***

The count?

7:

You know, the count. The number of Empire soldiers we saw today. While we were on patrol.

6:

Riiiiiiiiiight, “on patrol” *(doing finger quotes)*

7:

I saw three ships flying over while we were walking out of the ice cream— I mean… the nice green, the nice… green… building. Where were were definitely working. And doing what we were supposed to be doing.

3:

I saw two of them when we were wandering around in the— place we are totallly familiar with.

COMMANDER 1:

Excellent! So: how many soldiers is that altogether?

5:

Count those guys? How the heck are we supposed to do that? They’re all in ships! Flying around! In the air!

6:

Yeah, and the windows are super tiny!

7:

Commanders, with all due respect: we’re gonna need a little help here.

COMMANDER 1:

Didn’t any of you show up to Basic Training?

The others look sheepish

COMMANDER 2:

Ok, you knuckleheads. What Rebel Base are we at right now?

8:

Rebel… Base… Ten?

COMMANDER 1:

YES! Good job, number…

8:

Eight, commander. Number Eight.

COMMANDER 2:

Yes. Good job. We're at Rebel Base Ten. And conveniently, the Empirical Patrol Transports in this sector have a standard crew of HOW many people?

9:

...Ten?

COMMANDER 1:

YES! *(to 2 and also to self)* Thank goodness.

COMMANDER 2:

And the Empire follows the rules HOW often?

5:

Never?

6:

Sometimes?

4:

ALways.

COMMANDER 2:

Correct, Number Four. Always. SO: if you see an Empirical Patrol Transport in this sector, flying over Rebel Base Ten… How many people have to be inside?

3-9: ***(thinking it over, staring at their fingers)***

Umm… Ummm… Ummmmmmmmmm…

1 and 2 put hands on hips and lean pointedly towards them…

3-9: ***(in tiny voices)***

...Ten?

1 and COMMANDER 2:

YES!

3-9:

Woohoo! *(They high five, jump around, do secret handshakes, look real triumphant. They start to leave out the door the way they came in.)*

COMMANDER 1:

Hang on a minute!

COMMANDER 2:

Where do you think YOU're going?

3-9 freeze in their walking-out poses, and just swivel their heads in unison to look at the commanders.

COMMANDER 1:

What about the COUNT?

COMMANDER 2:

How many people total did you see today?

3-9 groan, walk back in, re-make their horizontal line in numerical order. They think some more.

5: ***(suddenly excited)***

Wait— hold on. I think I get it. Ooo, ooo! Patrol Group B: how many ships did you see at the ice cream shop?

7: ***(trying to play it off cool)***

Pshhhh what?

9:

That's totally not where we were.

8:

Two!

7 and 9 facepalm.

5:

Great! Two of you stand over there.

(Points towards stage right, so that the audience will eventually be able to read the number left to right. 8 pushes 7 and 9 over towards stage right where they stand together. 1 and 2 stand together watching, further off towards stage right)

6: *(also suddenly excited)*

I get it! Ok, Patrol Group A — how many ships did YOU see while you were lost?

3:

Three! I mean, we weren't—

6:

Yeah, ok ok ok… three of you go stand over there!

3 and 4 and 8 stand with 7 and 9, so there are five people in a clump

5: *(proud and excited)*

There— see?

7:

Uh… I'm pretty sure there were more than FIVE people, with all those ships flying around. Am I right? *(High-fives number 9)*

6: *(now on a roll, ready to bring it on home)*

Shhhhh! Did ANYone see any Empire people on the ground, NOT in a ship? (The rest of them think, and then shake their heads) OK! I'll be zero people!

6 Stands to Stage Left of the group of five, and makes a big 0 with their body/hands/somehow — so that the audience sees 5 people, and then a zero: ||||| 0

5:

Great! *(Studies their work)* Five… zero… five...zero… OK rebels, so if each ship has a crew of— how many?

8:

Ten!

5:

And we have five tens, so we have… *(concentrates REAL hard, and points like a conductor at each of the following people, who count off in turn, by tens.)*
They count off hesitatingly and then more confidently… this part, though it is just counting, should be the Most Dramatic Sequence Ever with epic highs and lows… background music? fireworks at the end? go nuts!)

3:

Ten.

4:

Twenty?

7:

Th—thirty!

8:

Forty?

9:

Fifty!

5: ***(almost as excited As Possible)***

Yes! And then: single people!

6: ***(triumphantly)***

Zero!

5: ***(wait now sudden danger)***

WAIT! I totally saw two of them playing frisbee in the park! (Runs over to join 6 who is still shaped like a zero; they look confused for a second and then they each hold up one finger to represent "one")

6: ***(Now AS excited As Possible)***

FROM THE TOP!

3:

Ten!

4:

Twenty!

7:

Thirty!

8:

Forty!

9:

Fifty!

5:

One!

6:

Two!

5 and 6:

Fifty-two!

3, 4, 7, 8, 9:

Fifty-two!

3-9:

FIFTY-TWO!!!!!

(They all start dancing around high-fiving and singing about fifty-two… fireworks? conga line? go nuts!)

COMMANDER 1: ***(eventually)***

Silence!

COMMANDER 2:

Fall in!

They all happily fall back in line, super delighted with themselves.

COMMANDER 1:

Fifty-two Empirical personnel.

COMMANDER 2:

Excellent work, rebels! *(Pats 5 and 6 on the back.)*

5 and 6:

Thank you, commander.

9:

Three cheers!

3, 4, 7, 8, 9:

Hip hip, hooray! Hip hip, hooray! Hip hip, hooray! *(They all applaud)*

COMMANDER 1: ***(cheerfully)***

Fifty-two. Fifty-two! Do you know what this means?

5 and 6:

(shaking their heads and grinning, maybe holding their hands out, expecting a prize)

COMMANDER 1: ***(suddenly angry)***

It means WE'RE BEING INVADED!

COMMANDER 2:

TO THE SHIP!

3-9 freak out and start running around panicking and/or getting ready

5:

Wait! *(They all freeze)* Where's Number Ten? Without Number Ten we can't fly a ship! We're just nine people on the ground!

10: ***(entering)***

Where have y'all been? I've been waiting for you in the ship!

3-9:

Woohoo! To the ship! *(They all exit)*

1 and 2 stay behind for a second, recovering from this roller-coaster.

COMMANDER 2:

It's all right, commander. Afterwards, I'll take you out for ice cream.

COMMANDER 1:

That would be nice, commander.

COMMANDER 2:

Shall we?

They exit.

END OF PLAY

USE THE ZERO, SEEKER!

by Mandy Khoshnevisan

MATH CONCEPTS:

Number place value up to hundreds; Using zero as a placeholder, to easily multiply numbers by 10; using zero as a number to multiply (equaling zero)

CHARACTERS:

SEEKER: a young spaceship pilot, eager to learn

BETA: a mysterious figure who lives in a swamp

SWAMP WALKERS: mysterious CREATURES who also live in a swamp and are probably dangerous. There are ten of them, but don't need to be played by ten actors (could use puppets or other props to play multiple characters each … if a two-person play, the swamp walkers could be played with puppets by Seeker and/or Beta)

PROPS:

Something to write on and something to write with (large enough for the audience to be able to see it)

SETTING:

Outdoors, on a forest planet

Lights up: A spaceship has landed on the forest planet. Someone climbs out of it.

SEEKER:

Hellooooooo? Hellloooooooo?? *(No answer. Looks around. There's nothing around.)* Great. THIS is the planet? THIS is where I'm supposed to go to find the Great Knowledge? *(Looks around some more)* I don't see anything but trees! Trees, trees, trees, as far as the eye can see. *(Looks at feet)* And muck. Trees and muck. Fantastic. I guess it's all a legend, and the great Beta isn't here after all.

SEEKER turns to leave; almost trips over Beta, who is a short and/or hunched over alien.

BETA:

Seeking me, are you?

SEEKER:

I'm not sure. Who ARE you?

BETA:

Beta I am. Beta is me. One plus one plus one is three.

SEEKER: ***(silently adds up one and one and one, nods.)***

Welllllll, that checks out. But how do I know that you're Beta?

BETA: ***(shrugs)***

You see anyone else here?

SEEKER:

Good point. Uh, okay, Beta, the great, the all-powerful, et cetera, et cetera,

BETA:

You seek the Great Knowledge, yes?

SEEKER:

Yes! What is it? I don't even know what it is! *(Kneels)* Tell me, great Beta, how do I obtain the Great Knowledge?

BETA:

Why are you wanting it?

SEEKER:

Uh… I mean… because it's Knowledge!

BETA:

Good enough for me. Let's get going, kid.

SEEKER:

Great! I mean, thank you! I mean, what do we do first?

BETA: ***(grandly)***

First… we must… complete your training.

SEEKER:

Okay, great! How long does that take?

BETA:

Months. Perhaps years!

SEEKER:

Aw, man! I don't have that much time!

BETA:

No one does. *(Sighing)* I guess we could take care of it with a montage.

Rockin' music starts to play, and/or, the other characters supply the music by singing some Eye of the Tiger-type wordless rockin' background music. Beta and Seeker do a montage of a bunch of different types of math training, possibly "wiped" or otherwise "scene-changed" by the other actors in the play... maybe Seeker is covering their eyes while Beta holds up different numbers on their hands for Seeker to guess… maybe Beta is counting Seeker's pushups or jumping jacks or other physical feats… maybe they're having a "naming numbers" battle where they're facing each other and just calling out random numbers… maybe Seeker is practicing counting as fast as they can while Beta times… etc etc… have fun, make up your own math montage!

Montage ends; back to the scene!

BETA:

You have done well, young one.

SEEKER: ***(panting)***

Thank you.

BETA:

I think it is time to give you … the Great Knowledge.

SEEKER:

Yes! Thank you! *(Kneels.)* I am ready!

BETA, with great ceremony, forms hands into a circle; Seeker, confused, copies them… pause.

BETA:

There it is…

SEEKER: ***(still holding hands in a circle)***

What? Where?

BETA:

In your hands, it is.

SEEKER: ***(breaks hands apart and looks in each of them.)***

Where? I don't see anything.

BETA:

Ah ah ahhhhh…. you see NOTHING.

SEEKER:

WHAT?

BETA:

Nothing. ZERO. (Holding up hands in a circle again) The Great Knowledge is ZERO.

SEEKER:

Oh boy oh boy. That's it. I'm outta here. *(Gets up to go)*

BETA:

Wait! You give up so easily? Zero has great power!

SEEKER gives a skeptical look

BETA:

True it is, my young friend! Watch!

Seeker comes back, reluctantly. With great ceremony, Beta gets out a whiteboard (or something else to write on)

BETA:

Say you have one sandwich *(draws a 1)* but you're still hungry. Just add zero *(draws in a zero)* and zippity zoom: you have TEN sandwiches!

SEEKER:

Keep talking.

BETA:

Then say you have a bunch of friends coming over, and they're hungry too! Just add aNOTHer zero *(draws another zero to the right of the "10" so that it reads "100")* and now you've got a hundred sandwiches!

SEEKER:

Let me try. *(Takes the whiteboard and erases it, or takes another piece of paper)* Okay, there's ONE of me *(draws a "1")* and then I want to start a band, so I add the power of ZERO *(draws a "0" on the wrong side, so that the board now reads "01")* and now there are ten of me. Right? *(Looks at Beta proudly)*

BETA:

Um, how shall I put this? No. Wrong.

SEEKER:

Whaddya mean? I put the magic zero on it, right?

BETA:

You can't just put the zero wherever you want! The right order it must be in! Otherwise no sense does it make!

SEEKER:

But I can understand YOU when you put WORDS in different places. Why not numbers?

BETA:

Numbers are DIFFerent. When you add a zero to the right-hand side, *multiplying by ten* you are. You moved the one into the ten's house. If you put a zero on the other side, like you just did, all you do is put a sign in the ten's house to show that there's no one in there.

SEEKER:

Hmmm. I get it, I think. *(Flips the paper/board upside down so that the "01" now reads "10")* How's that?

BETA: *(gasping)*

Truly you are the chosen one!

SEEKER:

Cool.

BETA:

Quick: multiply that by ten!

SEEKER:

Bam! (Draws another 0 so that now it reads "100") One hundred!

BETA:

Again!

SEEKER:

Boom! *(Draws another 0 so that now it reads "1000")* One thousand!

BETA:

Again!

SEEKER:

Pow! *(Draws another 0 so that now it reads "10000")* Um, … *(counts zeroes)*

BETA:

Ten thousands are …

SEEKER:

Oh! Right! Ten thousand! Does this just go on like this?

BETA:

Yes, as long as you keep multiplying by ten, you can just keep adding zeroes… until long after we have names for numbers …

SEEKER:

Wow… just like the stars!.......

They both look up into the sky for a beat or two, pondering the mysteries of the universe

SEEKER:

Wow… can you multiply anything by ten, just by adding a zero to the end?

BETA:

Darn tootin'! *(Erases board and/or gets another piece of paper; writes "56")* Fifty-six?

SEEKER:

Zap! *(Writes a zero so it reads "560")* Five hundred sixty!

BETA:

Yes! *(Writes "123")* One hundred twenty-three?

SEEKER:

Kapow! *(Writes a 0 so it reads "1230")* One thousand, two hundred thirty!

BETA:

Your final training challenge: .... this!

BETA writes a ridiculously long series of digits. SEEKER gets all dramatic and ready ... maybe more dramatic soundtrack here... finally, very theatrically, writes a 0 at the end of the string of numbers.

SEEKER:

Ta-daaaaa! It's ... THAT!

BETA:

Excellent! We'll learn how to say it in your next lesson. Congratulations, young one! You've learned your first trick with the power of Zero: how to multiply by ten!

Growling sounds nearby...

SEEKER: ***(a little scared)***

What's that?

BETA:

Oh... just a swamp walker.

SEEKER:

You don't seem worried.

BETA:

Frozen I am. This my terrified face is.

The swamp walker appears! It's totally scary. It growls at them.

SEEKER:

Should we run?!?

BETA:

I cannot.

SEEKER:

Then I will use my newfound power! *(Forms hands into a circle and "shoots" them at the MONSTER.)* Multiplyyyy byyyyy tennnnnn!

The MONSTER becomes ten MONSTERs. (This could be ten kids, or five kids with two puppets each, or some other fun solution.) The MONSTERs all growl and shuffle a little closer.

SEEKER:

Ack! There's ten of them!

BETA:

That's 'cause you multiplied them by ten!

SEEKER:

I panicked! Now what!

BETA:

I must teach you the second power… but quickly! Now, this is very dangerous.

SWAMP WALKERS:

Grrrowrrrrrrrrrrr!

SWAMP WALKERS *make a huge swipe at our heroes and take one big step closer. Our heroes back up one step. (SEEKER might have to drag BETA who is still very scared)*

BETA:

You must multiply by zero!

SEEKER:

What? How do I do that?

BETA:

Trust the power of Zero! Quick now!

SEEKER: ***(forms hands into circle, concentrates)***

Multiply by zero…

BETA:

Quick now! Use the Zero, Seeker!

SEEKER: ***(aiming the "zero" in their hands at the MONSTERs)***

MULTIPLYYYY BYYYY ZEROOOOOOOO!

The SWAMP WALKERS stop, growl surprisedly, look at themselves, look at each other, and, miming making sounds, dramatically fade away--completely silently.

SEEKER:

They're … gone!

BETA:

To multiply anything by zero … leaves zero.

SEEKER:

Anything?

BETA:

Anything.

SEEKER:

Wow. This IS some Great Knowledge.

BETA:

Told ya. And this is just the beginning.

They make "zeroes" with their hands and look up at the stars. Lights fade out.

END OF PLAY

GREATER THAN / LESS THAN

by Mandy Khoshnevisan

MATH CONCEPTS:

Greater than/less than: both the concept and the mathematical symbol

CHARACTERS:

HOTSHOT: A hotshot space pilot

PRACTICAL: Hotshot's friend

MONSTER: A hungry but confused CREATURE with a giant V-shaped mouth like a crocodile

CREATURES 1 through 4: Small animals, that on this planet are often prey

PROPS/NOTES:

Actor playing the MONSTER either holds their hands/jaws out like a sideways V (the greater-than/less-than symbol), or else carries something (rulers, etc) that form that sideways V-shape. The MONSTER could also be a puppet.

Also, you could have multiple kids playing the MONSTER at the same time so it's really long/large/scary; they could all speak the lines together, or just the head, or creatively switch them around

SETTING:

Outdoors in the wilderness on a strange planet

Lights up: our two heroes climb/crawl out of a crashed spaceship.

HOTSHOT:

Wow, that was kind of a bumpy landing.

PRACTICAL:

It was a crash.

HOTSHOT:

Potayto, potahto.

PRACTICAL:

That's the last time I let you fly me anywhere.

HOTSHOT:

Maybe the second to the last time; we gotta get outta here first.

PRACTICAL:

Yeah… where's here, anyway? Where are we? I thought you were taking me to the rebel base.

HOTSHOT:

Hey, come on, it's gotta be somewhere in this system. At least... no more than a few light years away, huh?

Ominous growl from offstage.

PRACTICAL:

Shh! Stop talking about yourself for a second.

HOTSHOT:

But I wasn't—

PRACTICAL:

Did you hear that?

HOTSHOT:

Hear what?

Another growl.

HOTSHOT:

Nope, I definitely did not hear that.

PRACTICAL:

Let's get out of here, shall we?

The MONSTER has entered where only HOTSHOT can see it.

HOTSHOT:

Too late.

They both turn and behold the beast.

MONSTER:

Hungry.

HOTSHOT: ***(trying to be polite so they can get away)***

Ex -excuse me?

MONSTER:

ME HUNGRY.

Four little CREATURES run in together in a group; the MONSTER has been chasing them, and hears or sees them enter.

MONSTER:

DINNER!

The MONSTER turns and faces the four CREATURES, so that its jaws are open towards them, and away from HOTSPUR and PRACTICAL — looks something like this, from the audience:) . . . . > o o

PRACTICAL:

Awww, they're so cute and scared! We should save them!

HOTSHOT:

YOU should save them. I'm outta here.

HOTSHOT tries to leave; PRACTICAL grabs them

PRACTICAL:

Not so fast, Hotshot. You got us into this mess. The least you can do is help SOMEone today.

HOTSHOT:

But look at its jaws! They're huge!

MONSTER:

SO HUNGRY. Must eat MANY THINGS!

CREATURES: ***(trembling and huddling together)***

Help us! eeeeeeeee!

PRACTICAL: ***(to HOTSHOT)***

Maybe we can distract it. *(To MONSTER)* Hey—uh, excuse me, MONSTER, sir, uh, ma'am, uh, your Monsterness? *(MONSTER turns towards them; the little CREATURES are still too scared to run away.)* Why don't you, uh, eat us instead? We're so much bigger.. Probably…?

MONSTER:

NOPE. Must eat the MOST THINGS. Four things greater than two things. WANT FOUR THINGS!

MONSTER turns back towards CREATURES and opens jaws wide.

CREATURES:

Eeeep! Help!

HOTSHOT:

Maybe we can fool it. *(whisper to CREATURES)* You, little CREATURES, squish together and we'll stand farther apart so we look bigger. *(they do so, and look real proud)*
(Louder, to MONSTER) Hey MONSTER! Look, there are more things on this side!

MONSTER looks at HOTSHOT and PRACTICAL and counts them to itself, then looks at the four CREATURES and counts them to itself, then looks back at our heroes.

MONSTER:

You think you fool ME? This MY planet. Me smart. You not more things. You just far apart! Far apart, still two things. Close together, still four things! Me big but me still can COUNT!

MONSTER advances on the CREATURES again, jaws wide in that sideways V so the audience can see that greater-than, less than look. CREATURES cower and try to show their tiny claws but they are too tiny and too scared.

CREATURE 1:

Back off, monster!

CREATURE 2:

We have claws—grrrr!

CREATURE 3:

We have teeth—om nom nom nom!

CREATURE 4:

It's no use, we're doomed!

All CREATURES:

Eeeeeeeeeek!

PRACTICAL:

Wait! I've got a plan… You! Little friend! Come here!

All CREATURES: ***(each pointing to themself)***

Me?

ALL CREATURES dash to the other side of the MONSTER to stand with HOTSHOT and PRACTICAL, so now there are six of them on one side.

MONSTER:

Wahoo! Easy pickings! One side zero, one side six! No need count, JUST EAT! *(MONSTER turns toward the group of six friends and opens jaws wide.)*

CREATURES:

EEEEEEEEK!

HOTSHOT:

Great plan.

PRACTICAL:

No, no, no, no. Let's try again. *(smooshes HOTSHOT and CREATURE 1 and 2 together)* YOU stay here. *(takes CREATURE 3 and 4 by the hand)* YOU come with me! *(they run to the empty side of the stage)*

HOTSHOT:

This better work!

The stage looks a little like this: .o. < .o. With the MONSTER in the middle

PRACTICAL:

Hey MONSTER! Count THIS!

MONSTER:

IS EASY. *(counts one side)* One, two, three! *(counts other side)* One, two …. *(looks terrified)* ...three.

CREATURES:

(gasp!)

MONSTER:

OH NO. No side have more things. Monster not know what do! *(MONSTER drops jaws, clutches head, looks distraught)*

CREATURES 1 & 2:

No side has more things!

CREATURES 3 & 4:

We're… equal!

HOTSHOT:

Guess we outsmarted you, monster!

MONSTER:

Me have rules. Must go now! *(starts to leave)*

HOTSHOT:

And good riddance, MONSTER!

MONSTER: ***(comes back in)***

Why need be so cruel? Monster have name! Monster name KEVIN!

PRACTICAL:

We didn't know!

MONSTER: ***(bursting into tears)***

You no ASK! *(MONSTER runs away crying)* Still so hungry!

CREATURES:

Thank you! Thank you! *(they do a victory dance of some kind)*

HOTSHOT:

Now we just have to find the rebel base.

CREATURE 1:

Base?

CREATURE 2:

We know base.

CREATURES 3 & 4:

Follow us!

PRACTICAL:

Well, Hotshot… I guess you didn't do so badly after all.

HOTSHOT:

Yep, that's right… I saved the day. I'm a hero.

They follow the bouncing CREATURES offstage.

END OF PLAY

THE SHAPE: EPISODE I

by Mandy Khoshnevisan

MATH CONCEPTS:

Three-dimensional shapes; volume; cubes; relative size; estimation

CHARACTERS:

COMMANDER 1: A Commander of the Rebel forces at Base 10 — the one in charge

COMMANDER 2: A Commander of the Rebel forces at Base 10 — second in command

3: one of the Rebels

4: one of the Rebels

PROPS/NOTES:

It might be nice to have "coffee" cups for the Commanders, and maybe a break room table and chairs for them to sit at, if that's available to you.

SETTING:

Rebel Base 10 (an outpost on a faraway planet)

Lights up: Commanders 1 and 2 are sitting in the break room, having coffee.

COMMANDER 1:

...And that's why I switched to decaf.

COMMANDER 2: ***(probably very bored)***

No kidding. Do you mind if we save that for the newsletter?

COMMANDER 1:

Go right ahead. *(with conviction)* My story deserves to be told!

3 and 4 rush in, panicked and panting

3 & 4:

Commander! Commander!

COMMANDER 1: ***(not standing up)***

At ease, ensigns.

3: ***(Saluting)***

I'm sorry, Commander. Were you in the middle of something?

COMMANDER 2: ***(relieved, standing up very quickly)***

No, no, Ensign. Not at all! Please. What seems to be the problem?

3 and 4 look at each other, not sure how to proceed.

3 & 4:

A SHAPE!

COMMANDERS 1 and 2 look at each other.

COMMANDERS 1 & 2:

A shape?

3 & 4:

A SHAPE!

COMMANDER 1:

What kind of shape?

3 and 4 look at each other.

3 & 4:

WE DON'T KNOW!

3:

Except for, it's really big.

4:

And it's moving towards us.

Commanders shake their heads disappointedly.

COMMANDER 1:

I'm afraid you're going to have to do better than that, Ensigns.

3 & 4:

Huh?

COMMANDER 1:

How big is big?

COMMANDER 2:

Moving towards us from where? And how fast?

3 and 4 hold their heads, look kinda panicky.

COMMANDER 1:

Let's start at the beginning: What KIND of shape is it?

COMMANDER 2:

Is it roundish? Is it squarish?

3:

It had *(thinks real hard)* fouuuur? Siiiides?

4:

Yes, four sides!

COMMANDER 1:

Hmmm… a quadrilateral.

COMMANDER 2:

That could be any number of things.

3 & 4:

Eeeeek!

COMMANDER 2:

Wait — is it flat? Or does it have volume?

3 & 4:

Volume?

COMMANDER 1:

You know, volume? *(using hand gestures to illustrate)* Is it flat like a piece of paper, or tall like a box?

3:

A box!

4:

Very much like a box!

COMMANDER 1:

Aha! Now we're getting somewhere.

COMMANDER 2:

Listen carefully: is it actually shaped like a box?

3 & 4:

Yes!

COMMANDER 1:

And the side of the shape you saw—

COMMANDER 2:

You mean the FACE—

COMMANDER 1:

Fine. The "face" of the shape you saw had four sides that were all about the same size?

3 & 4:

Yes!

COMMANDER 2:

And the other… faces? Same size?

3:

Looked that way, commander!

COMMANDER 1:

Aha! A CUBE!

COMMANDER 2:

Now we're getting somewhere.

3:

And alllll the faces looked real angry.

COMMANDER 2:

What do you mean, they looked angry?

COMMANDER 1: ***(now very exasperated)***

I'm talking about a "face," which is what you call a flat surface on a solid object *(maybe uses the surface of the table to help illustrate)* — not … *(gestures helplessly to own face)* … you know, a FACE!

4:

All the faces. They looked like this. *(3 and 4 make the scariest angriest faces they can think of)*

COMMANDER 1:

The faces had … FACES on them?

3:

Yeah. The faces. Like you said.

1 and 2 look at each other, now a little worried.

COMMANDER 1:

HOW big did you say this, uh, shape, was?

3 and 4 look around.

4:

Like, way bigger than this room.

3:

Like maybe… as big as … four of this room?

3 and 4 nod at each other in confirmation: they agree they are very good and smart.

COMMANDER 1:

Hmmmmmm. Why didn't you lead with THAT, ensigns?

3:

Huh? Isn't a giant shape scary enough for a report?

COMMANDER 1:

Yes, but "giant shape" could have been anything! An asteroid! A planet!

4:

A lollipop.

COMMANDER 1:

Yes, a lolli— NO! NOT a lollipop.

4:

A delicious, terrifying lollipop.

COMMANDER 1:

You idiots! An angry cube with six angry faces?

COMMANDER 2:

It's a Q-BOT! The largest one we've ever heard of!

COMMANDER 1:

Now, get the rest of the base together! We've got to go see what it wants! Maybe ONE of the faces will be willing to talk.

1 and 2 start to rush off. 3 and 4 stay put and clear their throats to get the Commanders' attention.

COMMANDER 2: ***(stopping near the door)***

What is it?

3:

Well, it's just that …

4:

Calling us idiots really hurt our feelings.

Commander 2 nudges Commander 1.

COMMANDER 1: ***(sighing)***

I'm SORRY, ok?

3:

I mean, just because someone doesn't know something doesn't make them an IDiot.

4:

Yeah. YEAH.

COMMANDER 2:

Look, friends. We all figured it out together, didn't we? Now, Commander, YOU learn to control your temper, and ensigns, YOU work on your descriptive language, and we'll avoid these kinds of misunderstandings in the future, all right? *(1 and 3 and 4 nod reluctantly.)* Now, everyone shake hands. *(They do)* Now, what say we all get together and confront that terrifying giant space robot?

They all pat each other on the back and stuff and start to leave.

COMMANDER 1:

Wait—where was it? Which direction?

3 and 4 look at each other, confused. Maybe they both point in different directions. 1 and 2 facepalm. Lights down.

END OF PLAY

THERE ARE MORE BASES OUT THERE. . .

by Mandy Khoshnevisan

MATH CONCEPT:
Counting in Base 10 versus other number bases (with base 6 as an example)

CHARACTERS:
At Rebel Base 10 —
5: Patrol squad A - not very smart common-sense-wise, but are unexpectedly good at math concepts.
6: Patrol squad B - not very smart common-sense-wise, but are unexpectedly good at math concepts.

From Rebel Base 6 —
Admiral Green 1
Ensign Green 5

PROPS/NOTES:
It's the break room, so there might be a table and chairs in there. There's also a trash can. The door and window can both be mimed. Also needed: something to write with and two pieces of paper (or something else to write on)

SETTING:
The break room of Rebel Base 10

Lights up: 5 and 6 are hanging out in the break room together.

5:
...And that's why unicorns are no longer allowed on our ships.

6:
NONE of our ships?

5:
Nope.

5 & 6: ***(disappointed)***

Awwwwww.

6:

Wait — what's that?

They both look out the "window," which is towards the audience.

5:

It's… looks like it's a crew from another base!

There is a knock at the door. (It is ENSIGN GREEN 5 from Base 6, but they don't know that yet.) 5 and 6 panic.

6: ***(calling to whoever's at the door)***

Who is it?

5: ***(also calling to whoever's at the door)***

We're definitely in charge, by the way!

6:

Yes, we're definitely allowed to be in here by ourselves.

ENSIGN GREEN 5:

Hello? Can you let me in, please?

5:

Ok. *(almost opens door)* WAIT. Who are you?

ENSIGN GREEN 5:

I'm from Rebel Base 6.

6:

Base 6? Never heard of ya.

ENSIGN GREEN 5:

If you open up, I'll explain.

6:

Ha! Yeah, right.

5:

If you're one of us: (says the first half of the secret password:) "I'm all out of mustard, can you help me out?"

ENSIGN GREEN 5: ***(sighing loudly)***

"I just brought guacamole, but I left the mustard in my other pants."

6:

Aha! I told you they were legit.

5:

No you didn't. *(Opens the door)*

ENSIGN GREEN 5 enters, relieved.

ENSIGN GREEN 5:

Thank goodness! I've been walking forever! Our ship ran out of fuel. Do you have any food scraps we could have, to get our ship in the air again?

6:

Sure thing! *(Goes to get the trash can)*

ENSIGN GREEN 5: ***(looking around)***

Wow, this place is pretty nice! You have so much space in here!

5:

Well yeah, ten people need a lot of room, right?

ENSIGN GREEN 5:

TEN people? No wonder! We only have six.

5 & 6:

SIX?

There is a knock on the door.

ENSIGN GREEN 5:

Uh oh. *(looks around for somewhere to hide)*

ADMIRAL 1: ***(from outside the door)***

Ensign?!? Are you in there?

ENSIGN GREEN 5: ***(sounding cheerful, but looking miserable)***

Admiral? Aye aye, Admiral! I'm in here, Admiral!

6: ***(impressed)***

Admiral? We don't have one of those.

ENSIGN GREEN 5:

Really? How do you get anything done? *(Opens the door)*

ADMIRAL 1:

Ensign?!? What's taking you so long?

ENSIGN GREEN 5:

I just got here! And then it took awhile for them to let me in.

5 and 6 get together and salute.

6:

We were just being overly cautious, Your Wonderful Admiral…ness.

ADMIRAL 1:

Well, well, well, friends. THIS is how many people are depending on you to get us back up in the air.... Let me write it down for you: *(takes a pice of paper, writes real big on it: "10" and holds it up so the audience can see)*

5:

Yes sir, Admiral, ma'am. Ten people.

ENSIGN GREEN 5:

Uh, no. Six.

5 & 6:

Ten!

ENSIGN GREEN 5:

Six!

5 & 6:

Ten!

ENSIGN GREEN 5:

Six!

ADMIRAL 1:

HOLD IT! *(They all freeze)* I see the problem. This is Rebel Base 10, isn't it? *(The other three nod vigorously)* Well, there you go. YOU see this and you see "ten," and YOU see this and you see "six," and you're both right. It just depends what base you're thinking about.

5 & 6 and ENSIGN GREEN 5:

Huh?

ADMIRAL 1:

Oh my stars. *(Flips over a new blank piece of paper on the pad)* OK, look. Right now, Base Ten friends, how many of YOU are there?

5 & 6: ***(looking at each other)***

Two.

ADMIRAL 1:

Correct! *(Writes down "2" on the paper)* See? Two.

ENSIGN GREEN 5:

I agree. That says two!

ADMIRAL 1:

(goes back to the original paper with "10" on it) Now, say you joined the rest of your team and got into the ship and took off, so that we could see ONE ship and NO people on the ground? *(Pointing to the 1 and then 0)* How many people is THAT?

5:

Hmmm, well, we need ten people to run one ship, so… TEN plus ZERO is..

6:

TEN!

ADMIRAL 1:

Correct! For you folks from base Ten, this says "Ten." BUT!

ENSIGN GREEN 5: ***(giggling under their breath)***

…butt!

ADMIRAL 1: ***(sharply)***

Ensign, how many of US are here right now? *(Gets the paper with 2 on it)*

ENSIGN GREEN 5: ***(counting)***

One…. two … Two, Admiral!

ADMIRAL 1:

Very good. *(Shows the paper with 2 on it, then grabs the paper with "10" also)* BUT!

ENSIGN GREEN 5 giggles again; ADMIRAL glares; GREEN 5 stifles it.

ADMIRAL 1: ***(continues)***

When we both go back to the ship, so we're all in ONE ship, with NONE of us on the ground… *(ADMIRAL points to the 1 and the 0)*

ENSIGN GREEN 5:

Well, Admiral… one of our ships needs exactly six people to fly.

ADMIRAL 1:

Yes, ensign… unfortunately I can't leave you behind.

ENSIGN GREEN 5:

Huh?

ADMIRAL 1:

Nothing.

ENSIGN GREEN 5: ***(looking at the sign with "10" on it)***

That makes … one ship plus no people equals SIX, Admiral.

ADMIRAL 1:

Correct. See?

The other three stare at the sign with "10" on it and go, "huh"

ADMIRAL 1:

Ten-hut! *(The others snap to attention automatically)*. Pop quiz!

5 & 6 and ENSIGN GREEN 5:

Awwww!

ADMIRAL 1:

Question 1 of 1! IF you two came to visit us at base six, and we flew above you in our ship, there would be ONE ship and TWO extra people which makes how many people? *(They add six and two together on their hands and together come to the answer)*

5 & 6 and ENSIGN GREEN 5:

EIGHT!

ADMIRAL 1:

And therefore, at base six, this number means… (takes the paper with a 2 on it and draws a 1 next to it so that it reads "12")

The others stare at it…

ENSIGN GREEN 5:

One ship...

5 & 6:

..plus two extra people…

5 & 6 and ENSIGN GREEN 5:

...equals EIGHT!

ADMIRAL 1:

Correct! *(Gives 5 & 6 the papers as if they were a prize)*

5 & 6 and ENSIGN GREEN 5: ***(staring at the papers and holding them up as if they were magical)***

Whoooooooooooa.

Like some sort of magical ritual and/or sports chant: the three Ensigns pass the paper with "10" on it to ENSIGN GREEN 5 (from Base 6), point at it, and say, "SIX!" Then they pass it to 5 and 6 (from Base 10) and say "TEN!" They pass it back - "SIX!" and back again - "TEN!" Then they take the paper with "12" on it and pass it to ENSIGN GREEN 5 and say, "EIGHT!" … back to 5 & 6 and say, "TWELVE!!!!!" back and forth "EIGHT!" "TWELVE!" … and so on… maybe this becomes a wild dance number or crazy riot…?

Eventually the ADMIRAL has had enough of this and successfully gets everyone to be quiet.

ADMIRAL 1:

Now, Ensign. Take the fuel and let's get going. We've got some other bases to check on before we get back home.

ENSIGN GREEN 5:

Yessir, Admiral, ma'am. *(To others)* Welp, nice to meet you!

(They exit with the garbage can, aka fuel.)

5 & 6 stare at the papers, deep in thought —then, struck belatedly by a thought, stare at the door. Then at each other.

5:

Other bases? OTHER other bases?

6:

Boy.... we gotta get out more.

END OF PLAY

THE SHAPE: EPISODE II

by Mandy Khoshnevisan

MATH CONCEPTS:

Two-dimensional shapes; estimating size and dimension; measurement units; multiplying numbers to determine area

CHARACTERS:

At Rebel Base 10 —
- COMMANDER 1: A Commander of the Rebel forces — the one in charge
- COMMANDER 2: A Commander of the Rebel forces — second in command
- 3: Patrol squad A
- 4: Patrol squad A

PROPS/NOTES:

Since we're in the break room, there might be chairs and a table, or something similar. The Ensigns also need a "SkyView," which can be anything with a space for one person to look through or into— a paper towel tube, a cup, a box, a tin can, a window... pretty much anything would do. They also need something to write with and something to write on.

You might also need either another person or people, or else some recorded sound, to make scary noises at certain times.

SETTING:

The break room of Rebel Base 10

Lights up: Commanders 1 and 2 are sitting in the break room, having coffee.

COMMANDER 1:

...And I said, "Penguin? I thought that was a guy who was late to a wedding!"

COMMANDER 2:

Ah, yes, I see. It's a joke. Very funny. Do you mind if I put that in the newsletter?

COMMANDER 1:

Go right ahead. Just give me credit, will you?

COMMANDER 2:

Absolutely, Commander.

3 and 4 rush in from outside, panicked and panting

3 and 4:

Commander! Commander!

COMMANDER 1: *(not standing up)*

At ease, ensigns.

3: *(Saluting)*

I'm sorry, Commander. Were you in the middle of something?

COMMANDER 2: *(relieved, standing up very quickly)*

No, no, Ensign. Not at all! Please. What seems to be the problem?

3 and 4 look at each other, not sure how to proceed.

3 and 4:

It's … A SHAPE!

COMMANDERS 1 and 2 look at each other.

COMMANDERS 1 and 2:

A shape?

3 and 4:

A SHAPE!

COMMANDER 1:

What kind of shape?

3 and 4 look at each other.

3 and 4:

WE DON'T KNOW!

COMMANDER 1:

This seems awfully familiar.

COMMANDER 2:

It's not the Q-BOT again, is it? We gave it all the cereal it wanted! Enough to last ages!

3:

No, no, Commander. It's ...

3 and 4:

A DIFFERENT SHAPE!

4:

It appeared on the ground right outside!

COMMANDER 2:

Should we investigate, Commander?

COMMANDER 1:

No, not yet. It might be dangerous. Think, ensigns. What did this shape look like?

3:

Green.

4:

Wiggly.

3:

Scary!

COMMANDER 1:

Wiggly, you say?

4:

Yes! It was all blobby, like … like some sort of blob!

COMMANDER 2:

Was it a blob?

3:

Sort of, commander… but it's not like, a big blobby blob. It's more like … a patch on the ground.

COMMANDER 1:

So it's flat?

COMMANDER 2:

Two-dimensional?

3:

Yes!

4:

A two-dimensional blob!

COMMANDER 1:

Hmmm….

COMMANDER 2:

Hmmmmmmmmm….

3 and 4: ***(not really knowing what the Commanders are thinking about, but wanting to feel like they're helping)***

HMMMMMmmmmmmmmmmmmmmmmmmmmmmm……

COMMANDER 1:

Ensigns, get out the SkyView.

3 and 4 go get the "SkyView," which can be anything with room for one kid to look through or into— a paper towel tube, a cup, a box, a tin can… pretty much anything would do.

3: (holding it up and/or ready to look into it)

Ready!

4: ***(grabbing it from 3 or pushing them out of the way)***

Ready!

COMMANDER 2:

Good! Now, set it up to look at the blob from above.

4 holds their space and refuses to let go, so 3 gives up and is the one to "calibrate" it in some way.

3 and 4:

Ready!

COMMANDER 1:

Can you see it? Can you see the shape?

4 looks, passes it back to 3 who also looks

3 and 4:

Yes, Commander!

COMMANDER 1:

NOW what does it look like? What's its approximate shape and size?

3 starts to hand it over directly to COMMANDER 1

4:

No, silly! Rebel Base regulations! Looking through the SkyView is an ensign's job! You know that!

3:

You're right. There has to be another way. *(thinks, helplessly)*

4: ***(after a beat, a sudden brainstorm!)***

Draw it!

They go get something to draw on -- a piece of paper, a whiteboard, etc. Upon it, they draw a rectangle with wiggly sides.

COMMANDER 1:

Aha! Now we're getting somewhere. Ensigns, how big is that shape?

3:

This big, commander. *(points to the paper)*

COMMANDER 1: ***(skeptical)***

Surely not!

3: ***(indignant)***

It is so! And don't call me Shirley!

COMMANDER 2: ***(breaking up a potentially tense situation)***

Ensign, stand down. Look, friends, this is just a *representation* of the shape. It's a picture. This is how big the *picture* is, but how big is the shape itself? *(pointing towards the outside)* Approximately, that is.

3 and 4:

Ohhhh.

4: ***(looking at the SkyView)***

Sayyyyyyyyyy, doesn't this thing have a measurement function on it?

3 and 4 take turns squinting through the SkyView and making invisible "pinching," as if they're trying to look at something and measure it with their fingers. When they're done, they write next to the sides of the wiggly rectangle how long they are: next to the little sides they write "10," and next to the big sides they write "15"

3:

This is approximate, commander.

4.

Approximate. Measuring all those wiggly sides would be too hard!

COMMANDER 1:

Understood. An estimate is all we need right now.

COMMANDER 2: ***(examining the measurements on the drawing)***

Commander—the small sides are equal to each other and the large sides are equal to each other -- that means it's rectangular!

COMMANDER 1:

That's great. Excellent work. Commander, go alert the rest of the squad.

(COMMANDER 2 turns to go.)

COMMANDER 1: ***(continued —with sudden urgency)***

Wait, Commander! *(Squinting at the drawing.)* Ten what? Fifteen what?

3:

Huh?

4:

What do you mean?

COMMANDER 1:

What are the units?

3:

Aren't we in the Unit?

4:

Rebel Base Ten Unit? *(they both salute)*

COMMANDER 1:

No… the UNITS. Ten inches? Ten nanometers? Ten feet? Ten MILES? Ten… bananas?

3:

Well, how long are the bananas?

COMMANDER 1:

EXACTLY!

3 and 4:

ohhhhh. (they confer, and peer through the SkyView) Ten FEET.

COMMANDER 1: ***(perhaps ready to rip out own hair in frustration)***

THANK you.

COMMANDER 2:

Commander, are you thinking what I'm thinking? *(1 looks like they're not sure yet but wants to say yes)* Well, if you WERE thinking what I'm thinking, you would be thinking, "Hmm, ten feet by fifteen feet, green oozy rectangular patch on the ground, wiggly sides … GHOST HOUSE!"

COMMANDER 1: ***(scared, then plays it off)***

EEEEEE! I mean…..YES, I was DEFinitely thinking that. Ghost house.

3 and 4:

Ghost house?!?

COMMANDER 2:

Apparently there was once a house there and it's trying to come back because it has unfinished house business.

3 and 4:

Oh, yeah. Makes sense.

COMMANDER 2:

We'll have to soak it up with something and scatter the pieces, quick, before we're haunted by a house forever!

COMMANDER 1:

Excellent work, Commander. What's the area? The square footage?

3: ***(scared)***

Quickly, please! I think I hear the Ghost House coming!

They all listen, hand to ear… there is an eerie sound of … ghost house. Is it a door creaking? Is it someone saying, "Iiiii'm a ghost houuuuuuuuuse and I'm gonna get youuuuuuuuuuu!" or someone saying "Live in meeeeee!" Choose your own funny ghost house sound.

They all react; some obviously scared, some scared but trying to pretend not to be.

4: ***(scoffing)***

Pssssh, the area? How are we supposed to figure THAT out?

COMMANDER 2:

Well, we'd have to multiply ten times fifteen to get the approximate area, and luckily a quick way to multiply something by ten is to take that zero from the ten and stick it to the end of the thing you're multiplying, so… 15 with a 0 on the end is … 150. A hundred and fifty square feet.

4: ***(deflated; impressed?)***

Oh. Like that.

(another ghost house noise but louder and scarier)

COMMANDER 1:

Squad! Think fast! What can we use to soak it up with? Paper towels?

COMMANDER 2:

Oh, I uh, used them all up.

COMMANDER 1:

You WHAT?

COMMANDER 2: ***(defensively)***

Well, I've been doing a lot of watercolor paintings lately.

COMMANDER 1:

After you used the last one, why didn't you buy more?!

COMMANDER 2:

I dunno, I've been busy!

The commanders start to argue some more… then there is a THIRD Ghost House noise that has the loudest and scariest sounds, and everyone freezes.

3:

…Commanders? *(Still frozen, the COMMANDERS just turn their heads to look at 3)* We've been ordering a lot of individual frozen pizzas lately, and I'd say the boxes are just about a square foot.

4:

How do you know that?

3:

They're square, and they're about a foot on each side. That's a square foot.

4:

Huh. Neat.

COMMANDER 1:

That's no good. We'd need a hundred and fifty of them. We have to cover the whole area.

4: ***(annoyed with 3 because of all the boxes)***

Oh, we probably have that many back there.

3: ***(breezily)***

Never throw anything away, that's what I say! You never know when it'll come in handy!

COMMANDER 1:

Brilliant work, Ensign! You may have saved us all. In fact, you deserve a medal. *(Scary GHOST HOUSE SOUND)* Quick, gather those boxes, and let's go get that ghost house! *(They do the thing where they all gather around and put a hand in)*. On three: Ghost house!

ALL:

One, two, three, GHOST HOUSE!

They put their hands in the air. COMMANDERS 1 and 2 rush off towards the back door where the trash cans are; 4 stops 3 from following.

4:

"Never throw anything away," huh? Isn't it your turn to take out the trash? And hasn't it been your turn for like... a month?

3: ***(shrugging)***

What can I say? When you got it, you got it.

4:

Ugh, typical. *(talking as they move to exit, the same direction the COMMANDERS went)* I'm gonna get you back. Somehow. You just wait.

3:

Not if I get you back first.

4: ***(pretty much offstage by now)***

For what?!?

(fade out, exiting and still arguing)

END OF PLAY

TEACHER PLANNING SECTION:

Here's a summary of the information that might be helpful in planning each play. This part doesn't need to be shared with students — especially the line counts, which are approximate and are just meant to be helpful for casting purposes in terms of which roles have the most (and the most complicated) speaking/reading needs. As we know, the character with the most to say is not by any means the most "important."

REBEL BASE 10

MATH CONCEPTS:

Estimating; number place value; counting in Base 10

CHARACTERS:

COMMANDER 1: *A Commander of the Rebel forces* — the one in charge. About 17 lines, some multi-sentence)

COMMANDER 2: *A Commander of the Rebel forces* — second in command About 16 lines, some multi-sentence

— characters 3-9 are all roles that would benefit from comic physicality

3: *on Patrol squad A* —	about 17 very short lines, some in unison with a group
4: *on Patrol squad A* —	about 10 very short lines, some in unison with a group
5: *on Patrol squad A* —	about 17 lines, some multi-sentence
6: *on Patrol squad B* —	about 13 lines, some multi-sentence
7: *on Patrol squad B* —	about 10 lines, some in unison but some multi-sentence
8: *on Patrol squad B* —	about 10 very short lines, some in unison with the group
9: *on Patrol squad B* —	about 15 very short lines, some in unison with the group
10: *The Pilot* —	1 line; the Pilot is offstage for almost the whole play.

PROPS/NOTES:

No props or furniture specifically called for; feel free to use your imagination!

SETTING:

We are in the meeting room of Rebel Base 10, somewhere a long while ago in a galaxy that's certainly not very close. There may well be Wars between Stars out there, is what I'm saying.

USE THE ZERO, SEEKER!

MATH CONCEPTS:

Number place value up to hundreds; using zero as a placeholder, to easily multiply numbers by 10; using zero as a number to multiply (and how anything multiplied by zero is zero)

CHARACTERS:

SEEKER: *a young spaceship pilot, eager to learn* — about 40 lines, many multi-sentence

BETA: *a mysterious figure who lives in a swamp* — about 45 lines, some shorter but many multi-sentence

SWAMP WALKERS: *mysterious CREATURES who also live in a swamp and are probably dangerous.* There are ten of them, but don't need to be played by ten individual actors (kids could use puppets or other props to play multiple characters each … if presented as a two-person play, the swamp walkers could be played with puppets by Seeker and/or Beta) — A group character; lots of physicality but no actual lines (apart from specific growling "lines")

PROPS:

Something to write on, and something to write with (large enough for the audience to be able to see it)

SETTING:

Outdoors, on a forest planet

OTHER NOTES:

GREATER THAN / LESS THAN

MATH CONCEPTS:
Greater than/less than: both the concept and the mathematical symbol

CHARACTERS:

HOTSHOT: *A hotshot space pilot* — about 18 lines, some multi-sentence

PRACTICAL: *Hotshot's friend* — about 15 lines, some multi-sentence

MONSTER: *A hungry but confused creature with a giant V-shaped mouth like a crocodile* — about 13 lines, many with multiple short sentences. Lots of physicality and comedy... could be played by one actor, or several, as needed.

CREATURES 1 through 4: *Small animals, that on this planet are often prey.* — Lots of physicality; each creature has about 8 lines: one line to itself and the rest of them spoken together with at least one other actor

PROPS/NOTES:

Actor playing the MONSTER either holds their hands/jaws out like a sideways V (the greater-than/less-than symbol), or else carries something (rulers, etc) that form that sideways V-shape. The MONSTER could also be a puppet.

Also, you could have multiple kids playing the MONSTER at the same time so it's really long/large/scary; they could all speak the lines together, or just the head, or creatively switch them around

SETTING:
Outdoors in the wilderness on a strange planet

OTHER NOTES:

THE SHAPE: EPISODE I

MATH CONCEPTS:
Three-dimensional shapes; volume; cubes; relative size; estimation

CHARACTERS:

COMMANDER 1: *A Commander of the Rebel forces at Base 10, the one in charge* — about 25 lines, a few multi-sentence

COMMANDER 2: *A Commander of the Rebel forces at Base 10, second in command* — about 14 lines, a few multi-sentence

3: *one of the Rebels* — about 15 short lines, some in unison.

4: *one of the Rebels* — about 18 short lines, some in unison

PROPS/NOTES:

It might be nice to have "coffee" cups for the Commanders, and maybe a break room table and chairs for them to sit at, if that's available to you.

SETTING:

In the break room of Rebel Base 10 (an outpost on a faraway planet)

OTHER NOTES:

THERE ARE MORE BASES OUT THERE...

MATH CONCEPTS:

Counting in Base 10 versus other number bases (with base 6 as an example)

CHARACTERS:

From Rebel Base 10 —

5: *on Patrol squad A; not very smart common-sense-wise, but unexpectedly good at math concepts.* — about 18 short lines, some in unison with others

6: *on Patrol squad B; not very smart common-sense-wise, but unexpectedly good at math concepts.* — about 18 short lines, some in unison with others

Visiting from Rebel Base 6 —

Admiral Green 1: *the one in charge at Rebel Base 6.* — about 20 lines, some multi-sentence.

Ensign Green 5: *a similar lower-level underling, just from another rebel base* — about 25 short lines, some in unison with others.

PROPS/NOTES:

It's the break room, so there might be a table and chairs in there. There's also a trash can. The door and window can both be mimed. Also needed: something to write with, and two pieces of paper (or something else) to write on.

SETTING:

The break room of Rebel Base 10

OTHER NOTES:

THE SHAPE: EPISODE II

MATH CONCEPTS:

Two-dimensional shapes; estimating size and dimension; measurement units; multiplying numbers to determine area

CHARACTERS:

At Rebel Base 10 —

COMMANDER 1: *A Commander of the Rebel forces — the one in charge* — about 28 lines, some multi-sentence

COMMANDER 2: *A Commander of the Rebel forces — second in command* — about 18 lines, many multi-sentence

3: on Patrol squad A — about 25 short lines, a few in unison

4: on Patrol squad A — about 30 short lines, a few in unison

PROPS/NOTES:

Since we're in the break room, there might be chairs and a table, or something similar.

The Ensigns also need a "SkyView," which can be anything with a space for one person to look through or into— a paper towel tube, a cup, a box, a tin can, a window… pretty much anything would do.

They also need something to write with, and something to write on.

You might also need either another person or people, or else some recorded sound, to make scary noises at certain times.

SETTING:

The break room of Rebel Base 10

OTHER NOTES:

Made in United States
North Haven, CT
05 November 2026

21767751R00037